Simplifying Civil Engineering Solutions With Peter Chew Rule, Method And Theorem

Peter Chew

PCET VENTURES (003368687-P)
Email:peterchew999@hotmail.my

Cover Design : Peter Chew
Cover Image: Freepik Premium

Author: Peter Chew

Peter Chew is Mathematician, Inventor and Biochemist. Global issue analyst, Reviewer for Europe Publisher, Engineering Mathcmatics Lccturcr and President of Research and Development Secondary School (IND) for Kedah State Association [2015-18].

Peter Chew received the Certificate of appreciation from Malaysian Health Minister Datuk Seri Dr. Adam Baba(2021), PSB Singapore. National QC Convention STAR AWARD (2 STAR), 2019 Outstanding Analyst Award from IMRF (International Multidisciplinary Research Foundation), IMFR Inventor Award 2020 , the Best Presentation Award at the 8th International Conference on Engineering Mathematics and

Physics ICEMP 2019 in Ningbo, China and Excellent award (Silver) of the virtual International, Invention, Innovation & Design Competition 2020 (3iDC2020).

Analytical articles published in local and international media. Author for 21 Book and 7 preprint articles published in the World Health Organization (WHO) .

Peter Chew also is CEO PCET, Ventures, Malaysia, PCET is a long research associate of IMRF (International Multidisciplinary Research Foundation), Institute of higher Education & Research with its HQ at India and Academic Chapters all over the world, PCET also Conference Partner in CoSMEd2021 by SEAMEO RECSAM.

Peter Chew as Keynote Speaker of the 8th International Conference on Computer Engineering and Mathematical Sciences (ICCEMS 2019) and the International Conference on Applications of Physics , Chemistry & Engineering Sciences, ICPCE 2020 . 2nd Plenary Speaker the 6th International Multidisciplinary Research Conference with a Mindanao Zonal Assembly on January 14, 2023, at the Immaculate Conception

University, Bajada Campus, Davao City. Keynote speaker for Scholars World Congress on Nanomedicine and Advanced Drug Delivery. 10-11 Jul 23. Paris, France. THEME: "Frontiers in Nanomedicine and Advanced Drug Delivery Systems" by Drug Delivery,

Special Talk Speaker at the 2019 International Conference on Advances in Mathematics, Statistics and Computer Science, the 100th CONF of the IMRF,2019, Goa , India.

Invite Speaker of the 24th Asian Mathematical Technology Conference (ATCM 2019) Leshan China and the 5th(2020), 6th (2021) and 7th (2022) International Conference on Management, Engineering, Science, Social Sciences and Humanities by Society For Research Development(SRD).

Peter Chew is also Program Chair for the 11th International Conference on Engineering Mathematics and Physics (ICEMP 2022, Saint-Étienne, France | July 7-9, 2022) and Program Chair for the 12th International Conference on Engineering Mathematics and Physics (ICEMP 2023, Kuala Lumpur, Malaysia | July 5-7, 2023). For more information, please get it from this link Orcid: https://orcid.org/0000-0002-5935-3041.

SIMPLIFYING CIVIL ENGINEERING SOLUTIONS WITH PETER CHEW RULE, METHOD AND THEOREM

TABLE OF CONTENTS

TABLE OF CONTENTS

"Simplifying Civil Engineering Solutions With Peter Chew Rule , Method And Theorem"

Civil Engineering has always been a challenging field, requiring engineers to solve some complex problems. The Peter Chew Rule, Method, and Theorem are key tools that enable engineers to simplify solve some complex engineering problems.

The purpose Peter Chew Rule for solution of triangle is to provide a simple method compare current methods to aid in mathematics teaching and learning. By using the Peter Chew Rule, we can solve some Engineering problems more accurately and with greater ease. This can be especially helpful for students and educators in the field of Mathematics and Engineering.

Peter Chew's rule is a better way to calculate certain Engineering Mathematics problem than using cosine rule methods that involve taking the square root step. This is because the square root step can sometimes be imprecise, leading to less accurate results overall. By using Peter Chew's rule, which does not require this step, you can get more accurate answers that will give you a better understanding of what you're measuring or calculating. So if you want to make sure your answers are as accurate as possible, consider using Peter Chew's rule instead of using cosine rule methods that may not be as precise.

Peter Chew Method for solving triangles problem was developed with the goal of providing a simple approach to aid in teaching and learning mathematics. By applying this method to some Engineering problems, we can make the learning of Engineering more accessible and less daunting for students.

Peter Chew's theorem is a valuable tool in the age of Artificial Intelligence, as it can be used to convert all Quadratic Surds more easily and quickly than current methods. This can greatly improve the effectiveness of teaching and learning mathematics. In the case of future epidemics such as Covid-19, when students may have to study from home, Peter Chew's theorem can help facilitate remote mathematics education.

Presenting numbers in *surd* form is quite common in science and engineering especially where a calculator is either not allowed or unavailable, and the calculations to be undertaken involve irrational values. In these cases, the Peter Chew theorem can be a valuable tool in teaching and learning engineering, as it allows for exact value answers to be achieved.

Peter Chew Rule and Method are Simple Solution in Peter Chew Triangle Diagram and Peter Chew Triangle Diagram Peter Chew Triangle Diagram has passed double-blind review by The 12th International Conference on Engineering Mathematics and Physics, ICEMP 2023. Peter Chew Triangle Diagram(preprint) is share at World Health Organization(WHO_

Peter Chew Theorem for Quadratic Surds also has passed double-blind review by The 12th International Conference on Engineering Mathematics and Physics, ICEMP 2023. Peter Chew Theorem (preprint) is also share at World Health Organization

Peter Chew [https://orcid.org/0000-0002-5935-3041]

Mathematician , Inventor and Biochemist.

Peter Chew Rule and Method are simple solution in Peter Chew Triangle Diagram and Peter Chew Triangle Diagram Peter Chew Triangle Diagram has passed double-blind review by The 12th International Conference on Engineering Mathematics and Physics, ICEMP 2023. Peter Chew Triangle Diagram will be collected in Journal of Physics: Conference Series (doi:10.1088/issn.1742-6596 ; Online ISSN: 1742-6596 / Print ISSN: 1742-6588). Conference Proceedings Citation Index–Science(CPCI-S)(Thomson Reuters, WoS), Scopus, Ei Compendex, Inspec(IET), etc.

2023 The 12th International Conference on Engineering Mathematics and Physics
Kuala Lumpur, Malaysia | July 5-7, 2023

Acceptance Letter

(Full Paper for Publication and Presentation)

Paper ID CP003
Paper Title Peter Chew Triangle Diagram And Application
Author(s) Peter Chew

Registration Deadline
April 30, 2023

To whom it may concern,

Congratulations on the acceptance of your paper! And thank you for your interest in 2023 The 12th International Conference on Engineering Mathematics and Physics (http://www.icemp.org/).

Now we are pleased to inform you that your paper identified above has been accepted for presentation and publication at the conference after the strict reviewing process. And you are cordially invited to present your paper at ICEMP 2023. **And after registration and presentation, your paper will be collected in Journal of Physics: Conference Series (doi:10.1088 /issn.1742-6596; Online ISSN: 1742-6596 / Print ISSN: 1742-6588).** The registration procedure is in next page.

Peter Chew Theorem for Quadratic Surds has passed double-blind review by The 12th International Conference on Engineering Mathematics and Physics, ICEMP 2023.

2023 The 12th International Conference on Engineering Mathematics and Physics
Kuala Lumpur, Malaysia | July 5-7, 2023

Acceptance Letter

(Full Paper for Publication and Presentation)

Paper ID	CP009	**Registration Deadline**
Paper Title	Peter Chew Theorem and Application	**April 30, 2023**
Author(s)	Peter Chew	

To whom it may concern,

Congratulations on the acceptance of your paper! And thank you for your interest in 2023 The 12th International Conference on Engineering Mathematics and Physics (http://www.icemp.org/).

Now we are pleased to inform you that your paper identified above has been accepted for presentation and publication at the conference after the strict reviewing process. And you are cordially invited to present your paper at ICEMP 2023. **And after registration and presentation, your paper will be collected in Journal of Physics: Conference Series (doi:10.1088/issn.1742-6596; Online ISSN: 1742-6596 / Print ISSN: 1742-6588).** The registration procedure is in next page.

Peter Chew Triangle Diagram(preprint) is share at World Health Organization because the purpose of Peter Chew Triangle Diagram is to help teaching mathematics more easily , especially when similar covid-19 problems arise in the future. https://pesquisa.bvsalud.org/global-literature-on-novel-coronavirus-2019-ncov/resource/en/ppzbmed-10.20944.preprints202106.0221.v1

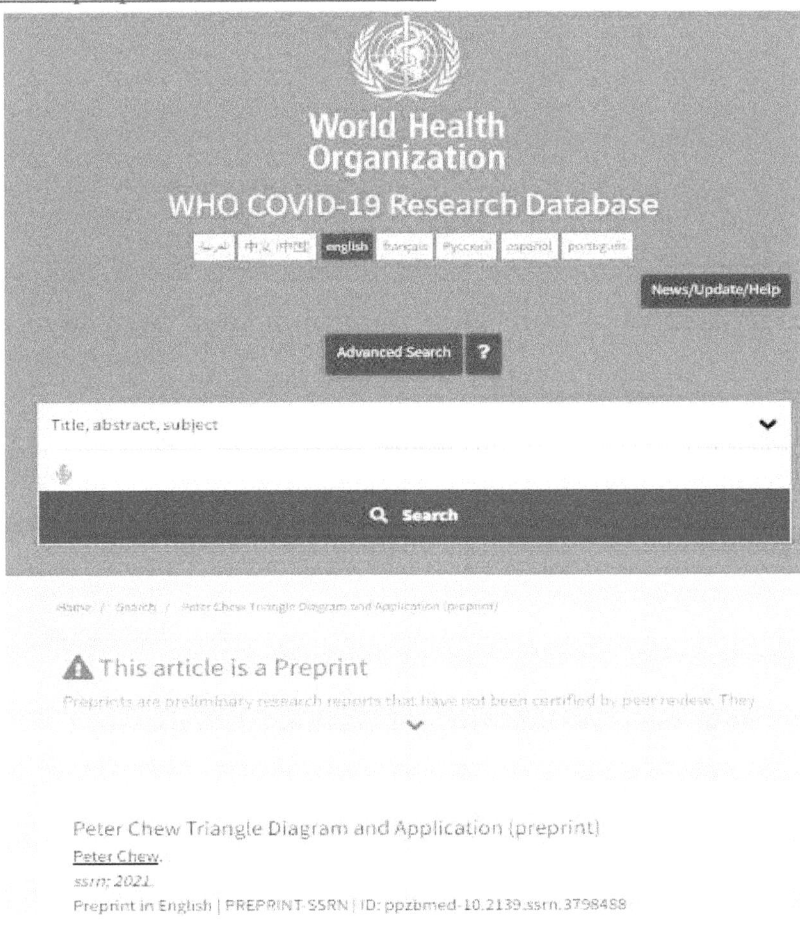

•

Peter Chew Theorem and Application (preprint) is share at World Health Organization because the purpose of Peter Chew Theorem is to help teaching mathematics, especially when similar covid-19 problems arise in the future.

https://pesquisa.bvsalud.org/global-literature-on-novel-coronavirus-2019-ncov/resource/en/ppzbmed-10.2139.ssrn.3798498

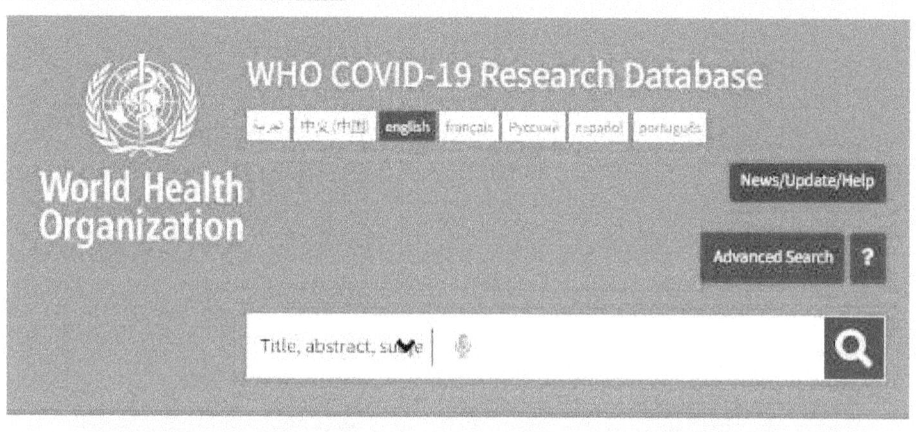

Home / Search / Peter Chew Theorem and Application (preprint)

This article is a Preprint

Preprints are preliminary research reports that have not been certified by

Fulltext

Print

XML

Search on Google

Peter Chew Theorem and Application (preprint)

Peter Chew.

ssrn; 2021.

Preprint in English | PREPRINT-SSRN | ID: ppzbmed-10.2139.ssrn.3798498

Full text: Available
Collection: Preprints
Database: PREPRINT-SSRN
Main subject: COVID-19
Language: English
Year: 2021
Document Type: Preprint

ABSTRACT

The Objective of Peter Chew Theorem is to make it

Chapter 1 : Application of Peter Chew Rule in Civil Engineering

Abstract.

The purpose of Peter Chew Rule for solution of triangle[1] is to provide a simple method to compare current methods to aid in mathematics teaching and learning, especially if similar COVID-19 problems arise in the future. Therefore, applying the Peter Chew's Rule for solution of triangle in civil engineering can make teaching and learning of civil engineering easier.

Besides being simple, Peter Chew rule is also more accurate than current methods, because the main advantage of Peter Chew's rule is that there is no square root step. Usually the square root value is only an approximation. Therefore, using Peter Chew's rule will give a more accurate answer than using the cosine rule involving a square root step.

The purpose of Peter Chew's Rule for solution of triangle is the same as Albert Einstein's famous quote Everything should be made as simple as possible, but not simpler.

Keywords: Civil Engineering, Application Peter Chew Rule, solution of triangle..

1. Introduction, The Leaning Tower.

1.1_ Leaning Tower of Pisa

The Leaning Tower of Pisa[2] (Italian: *torre pendente di Pisa*), or simply the Tower of Pisa (*torre di Pisa* [ˈtorre di ˈpiːza; ˈpiːsa][3]), is the *campanile*, or freestanding bell tower, of the cathedral of the Italian city of Pisa, known worldwide for its nearly four-degree lean, the result of an unstable foundation.

The tower is situated behind the Pisa Cathedral and is the third-oldest structure in the city's Cathedral Square (*Piazza del Duomo*), after the cathedral and the Pisa Baptistry.

The height of the tower is 55.86 metres (183 feet 3 inches) from the ground on the low side and 56.67 m (185 ft 11 in) on the high side. The width of the walls at the base is 2.44 m (8 ft 0 in).

Its weight is estimated at 14,500 tonnes (16,000 short tons).[4] The tower has 296 or 294 steps; the seventh floor has two fewer steps on the north-facing staircase.

The tower began to lean during construction in the 12th century, due to soft ground which could not properly support the structure's weight, and it worsened through the completion of construction in the 14th century.

By 1990, the tilt had reached 5.5 degrees.[5,6,7] The structure was stabilized by remedial work between 1993 and 2001, which reduced the tilt to 3.97 degrees.[8]

1.2 Capital Gate

Capital Gate[9], also known as the Leaning Tower of Abu Dhabi, is a skyscraper in Abu Dhabi that is over 160 meters (520 ft) tall, 35 stories high, with over 16,000 square meters (170,000 sq ft) of usable office space.[10]

Capital Gate is one of the tallest buildings in the city and was designed to incline 18° west.[11] The building is owned and was developed by the Abu Dhabi National Exhibitions Company. The tower is the focal point of Capital Centre.

Foundation

The structure rests on a foundation of 490 pilings that have been drilled 30 meters (98 ft) below ground. The deep pilings provide stability against strong winds, gravitational pull, and seismic pressures that arise due to the incline of the building.

Of the 490 pilings, 287 are 1 meter (3 ft 3 in) in diameter and 20 to 30 meters (66 to 98 ft) deep, and 203 are 60 centimeters (24 in) in diameter and 20 meters (66 ft) deep. All 490 piles are capped together using a densely reinforced concrete mat footing nearly 2 meters (6.6 ft) deep. Some of the piles were only initially compressed during construction to support the lower floors of the building. Now they are in tension as additional stress caused by the overhang has been applied.[12]

Architecture and design

The building has a diagrid specially designed to absorb and channel the forces created by wind and seismic loading, as well as the gradient of Capital Gate. Capital Gate is one of only a handful of diagrid buildings in the world. Others include London's 30 St Mary Axe (Gherkin), New York's Hearst Tower, and Beijing's National Stadium.

Capital Gate was designed by architectural firm RMJM and was completed in 2011. The tower includes 16,000 square meters (170,000 sq ft) of office space and the Andaz Hotel on floors 18 through 33.[13,14]

2. Application of Peter Chew Rule for Solution Of Triangle in Civil Engineering

The degree of inclination the Leaning Tower

Example 1: The elevation angles of a leaning tower top measure from two point D and B on ground level are 40° and 50° respectively. Given that that DB = 3m, BC= 4m, find the degree of inclination of the leaning tower. Figure below.

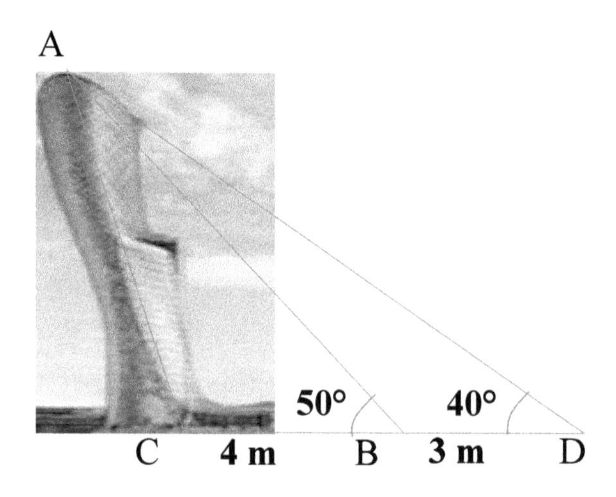

What's now. $\angle BAD = 50° - 40° = 10°$

$$\frac{3}{sin10°} = \frac{AB}{\sin 40°}$$

$$AB = 11.105 \text{ m}$$

Current Solution

Cosine rule, $AC^2 = a^2 + c^2 - 2ac\ cos\angle B$

$AC^2 = 4^2 + 11.105^2$

$-2(4)(11.105)\cos 50°$

$= 82.2158$

AC = 9.0673 m

Sine Rule, $\frac{11.105}{sin\angle ACB} = \frac{9.0673}{\sin 50°}$

$\sin\angle ACB = 0.9382$

$\angle ACB = 110.2486°, 69.7514°$ (reject)

The degree of inclination tower is 110.2486°.

Peter Chew Rule For Solution Of Triangle

$\tan\angle C = \frac{11.105 \sin\angle 50°}{4 - 11.105 \cos\angle 50°}$

$= - 2.7108$

$\angle C = 110.2488°$

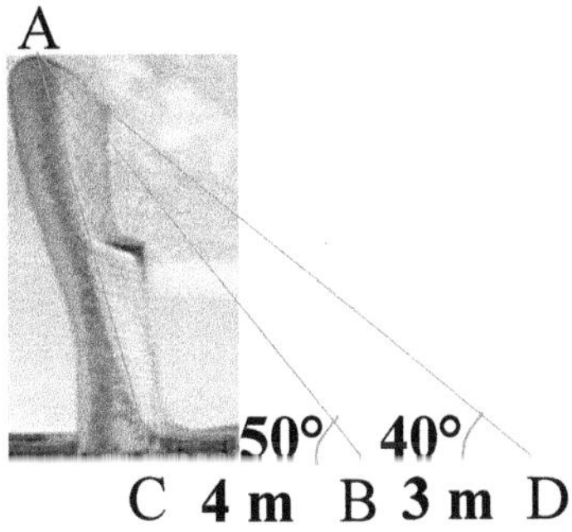

A

50° 40°

C **4 m** B **3 m** D

The degree of inclination of the leaning tower is 110.2488°.

Example 2: The elevation angles of a leaning tower top measure from two point D and B on ground level are 50° and 70° respectively. Given that that DB = 2 m , BC= 5 m, find the degree of inclination of the leaning tower

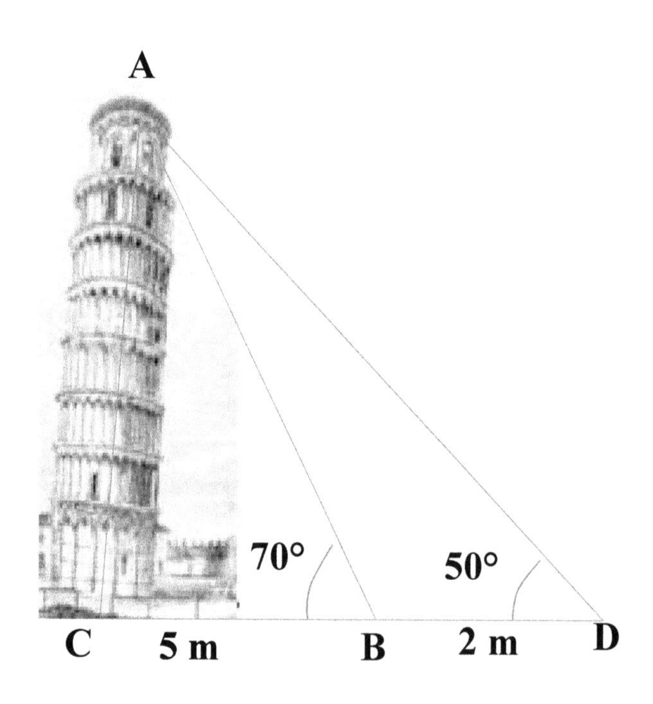

$\angle$BAD=70° − 50° = 20° .

$$\frac{2}{sin20°} = \frac{AB}{\sin 70°}$$

AB = 5.495 m

Current Solution

Cosine rule, $AC^2 = a^2 + c^2 - 2ac\ cos\angle B$

$AC^2 = 5^2 + 5.495^2$

$-2(5)(5.495)\cos 70°$

$= 36.401$

AC = 6.0333 m

Sine Rule, $\frac{5.495}{sin\angle ACB} = \frac{6.0333}{\sin 70°}$

$\sin\angle ACB = 0.85585$

$\angle ACB = 58.8538°, 121.1462°$ (reject)

The degree of inclination of the leaning tower is 58.8538°.

Peter Chew Rule For Solution Of Triangle

$\tan\angle C = \frac{5.495\sin 70°}{5-5.495\cos 70°}$

$= 1.6547$

$\angle C = 58.8538°$

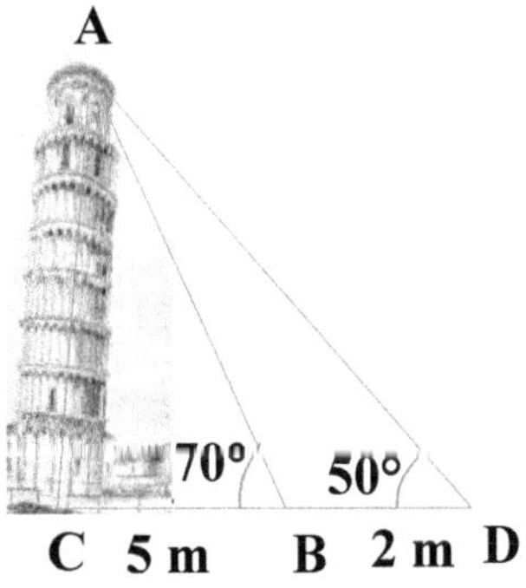

A

70° 50°

C 5 m B 2 m D

The degree of inclination of the leaning tower is 58.8538°.

3. Conclusion

The purpose Peter Chew **Rule for solution of triangle** is to provide a simple method compare current methods to aid in mathematics teaching and learning, especially if similar COVID-19 problems arise in the future.

Therefore, applying Peter Chew rule to find t**he** degree of inclination the Leaning Tower (Civil Engineering) problems can help to solve t**he** degree of inclination the Leaning Tower (Civil Engineering) problems more easily .

Besides being simple, Peter Chew's rule is also more accurate than current methods, because the main advantage of Peter Chew's rule is that there is no square root step. Usually the square root value is only an approximation.

The purpose of Peter Chew Rule for solution of triangle is the same as Albert Einstein's famous quote **Everything should be made as simple as possible, but not simpler.**

In addition, Albert Einstein's also quote :

i) We cannot solve our problems with the same thinking we used when we created them

ii) If you can't explain it simply you don't understand it well enough,

iii) "Genius is making complex ideas simple, not making simple ideas complex.".

iv). "Any intelligent fool can make things bigger and more complex. It takes a touch of genius - and a lot of courage - to move in the opposite direction."

*v)*God always takes the simplest way.

vi)When the solution is simple, God is answering.

Isaac Newton quote Nature is pleased with simplicity. And nature is no dummy.

From the Albert Einstein's and Isaac Newton quote above, it can be seen that simplifying knowledge is very important.

4. Reference

[1]. Chew, Peter, Peter Chew Rule for Solution of Triangle (2019). 2019 the 8th International Conference on Engineering Mathematics and Physics, Journal of Physics: Conference Series1411 (2019) 012009, IOP Publishing, doi:10.1088/1742-6596/1411/1/012009, Available at SSRN: https://ssrn.com/abstract=3843433

2. Leaning Tower of Pisa. Wikipedia, the free encyclopedia.

https://en.wikipedia.org/wiki/Leaning_Tower_of_Pisa

3. "DiPI Online". *Dizionario di Pronuncia Italiana (in Italian).* Archived *from the original on 30 October 2020.* Retrieved 26 December 2020.

4. "Leaning Tower of Pisa Facts". *Leaning Tower of Pisa.* Archived *from the original on 11 September 2013.* Retrieved 5 October 2013.

5. "Europe | Saving the Leaning Tower". *BBC News. 15 December 2001.* Archived *from the original on 21 September 2013*. Retrieved 9 May 2009.^

6. "Tower of Pisa". *Archidose.org. 17 June 2001. Archived from* the original *on 26 June 2009*. Retrieved 9 May 2009.

7. "Leaning Tower of Pisa (tower, Pisa, Italy) – Britannica Online Encyclopedia". *Britannica.com.* Archived *from the original on 8 March 2013*. Retrieved 9 May 2009.

8. "Leaning tower of Pisa loses crooked crown". *Irish News.* Archived *from the original on 28 November 2020.* Retrieved 10 June 2020.

9. Capital Gate. Wikipedia, the free encyclopedia. https://en.wikipedia.org/wiki/Capital_Gate

10. "Capital Gate / RMJM". *ArchDaily. 2018-04-28.* Archived *from the original on 2018-07-22.* Retrieved 2018-09-27.

11. "Capital Gate". *Abu Dhabi National Exhibitions Company (ADNEC). 2010. Archived from* the original *on 11 June 2010*. Retrieved 7 June 2010.

12. Mace Group, http://www.macegroup.com/media-centre/advanced-diagrid-technology-gives-shape-to-capital-gate Archived 2015-10-01 at the Wayback Machine | retrieved=July 29, 2015

13. "Backgrounder - Capital Gate Abu Dhabi". *Hyatt Hotels*. Archived *from the original on 2018-09-09.*

14. Capital Gate Atlas Obscura

(www.atlasosbcura.com). Retrieved on 2019-08-04.

Chapter 2 : Application of Peter Chew Method in Civil Engineering

Abstract.

The purpose of Peter Chew's method for solution of triangle[1] is to provide **a simple method** to compare current methods to aid in mathematics teaching and learning, especially if similar COVID-19 problems arise in the future.

Therefore, applying the Peter Chew's method for solution of triangle in civil engineering can make teaching and learning of civil engineering easier.

The purpose of Peter Chew's method for solution of triangle is the same as **Albert Einstein's famous quote Everything should be made as simple as possible, but not simpler.**

Kcywords: Civil Enginccring, Application Pctcr Chcw Mcthod, solution of triangle..

1. Introduction, the Leaning Tower.

1.1_ Leaning Tower of Pisa

The Leaning Tower of Pisa[2] (Italian: *torre pendente di Pisa*), or simply the Tower of Pisa (*torre di Pisa* [ˈtorre di ˈpiːza; ˈpiːsa][3]), is the *campanile*, or freestanding bell tower, of the cathedral of the Italian city of Pisa, known worldwide for its nearly four-degree lean, the result of an unstable foundation.

The tower is situated behind the Pisa Cathedral and is the third-oldest structure in the city's Cathedral Square (*Piazza del Duomo*), after the cathedral and the Pisa Baptistry.

The height of the tower is 55.86 metres (183 feet 3 inches) from the ground on the low side and 56.67 m (185 ft 11 in) on the high side. The width of the walls at the base is 2.44 m (8 ft 0 in).

Its weight is estimated at 14,500 tonnes (16,000 short tons).[4] The tower has 296 or 294 steps; the seventh floor has two fewer steps on the north-facing staircase.

The tower began to lean during construction in the 12th century, due to soft ground which could not properly support the structure's weight, and it worsened through the completion of construction in the 14th century.

By 1990, the tilt had reached 5.5 degrees.[5,6,7] The structure was stabilized by remedial work between 1993 and 2001, which reduced the tilt to 3.97 degrees.[8]

1.2 Capital Gate

Capital Gate[9], also known as the Leaning Tower of Abu Dhabi, is a skyscraper in Abu Dhabi that is over 160 meters (520 ft) tall, 35 stories high, with over 16,000 square meters (170,000 sq ft) of usable office space.[10]

Capital Gate is one of the tallest buildings in the city and was designed to incline 18° west.[11] The building is owned and was developed by the Abu Dhabi National Exhibitions Company. The tower is the focal point of Capital Centre.

Foundation

The structure rests on a foundation of 490 pilings that have been drilled 30 meters (98 ft) below ground. The deep pilings provide stability against strong winds, gravitational pull, and seismic pressures that arise due to the incline of the building. Of the 490 pilings, 287 are 1 meter (3 ft 3 in) in diameter and 20 to 30 meters (66 to 98 ft) deep, and 203 are 60 centimeters (24 in) in diameter and 20 meters (66 ft) deep.

All 490 piles are capped together using a densely reinforced concrete mat footing nearly 2 meters (6.6 ft) deep. Some of the piles were only initially compressed during construction to support the lower floors of the building. Now they are in tension as additional stress caused by the overhang has been applied.[12]

Architecture and design

The building has a diagrid specially designed to absorb and channel the forces created by wind and seismic loading, as well as the gradient of Capital Gate.

Capital Gate is one of only a handful of diagrid buildings in the world. Others include London's 30 St Mary Axe (Gherkin), New York's Hearst Tower, and Beijing's National Stadium.[12]

Capital Gate was designed by architectural firm RMJM and was completed in 2011. The tower includes 16,000 square meters (170,000 sq ft) of office space and the Andaz Hotel on floors 18 through 33.[13,14]

2. Current Method and Peter Chew Method for solution of triangle.

Example, Find third side, b

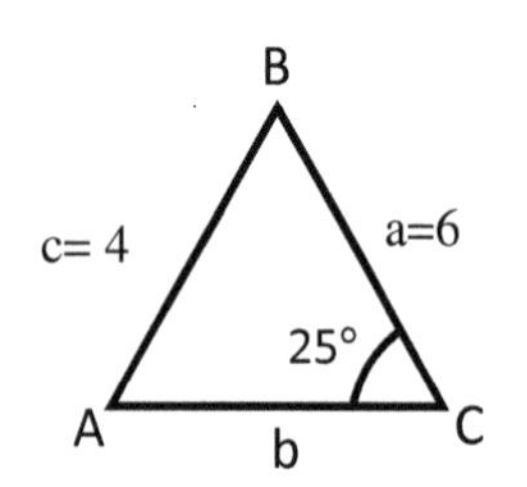

Current Method

Step 1:

$$\frac{6}{\sin A} = \frac{4}{\sin (25°)}$$

Rearranging gives us:

$$\sin A = \frac{6 \cdot \sin (25°)}{4}$$

$$\sin A = 0.6339$$

$$A = 39.34° , 140.66°$$

,

Step 2:

When A = 39.34° , $B = 180° - 39.34° - 25° = 115.66°$

When A = 140.66° , $B = 180° - 140.66° - 25° = 14.34°$

Step 3:

When B = 115.66°

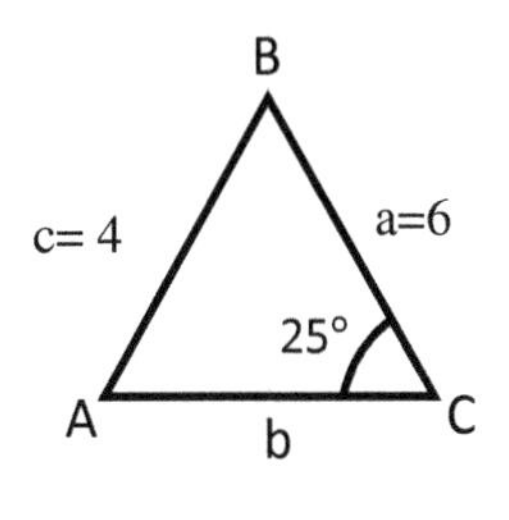

$$\frac{b}{\sin 115.66^\circ} = \frac{4}{\sin 25^\circ}$$

$$b = \frac{4 \ sin\ 115.66^\circ}{sin\ 25^\circ}$$

$$= 8.531$$

When B = 14.34°

$$\frac{b}{\sin 14.34^\circ} = \frac{4}{\sin 25^\circ}$$

$$b = \frac{4 \ sin\ 14.34^\circ}{sin\ 25^\circ}$$

$$= 2.344$$

$\therefore$ b = 8.531 , 2.344

Peter Chew Method: Using cosine rule

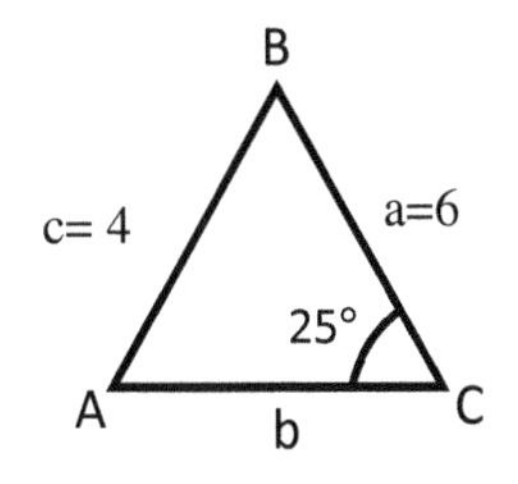

$$4^2 = b^2 + 6^2 - 2(b)(6)\ cos25^\circ$$

$$b^2 - 10.8757\ b + 20 = 0$$

b = 8.531 , 2.344

3. Using Math portal [15] for solving triangle problem

[Calculated on Feb 1, 2023]

MathPortal website's owner is mathematician Miloš Petrović.

Example: Find side a of a triangle if side b=3, side c = 5 and angle C = 35° .

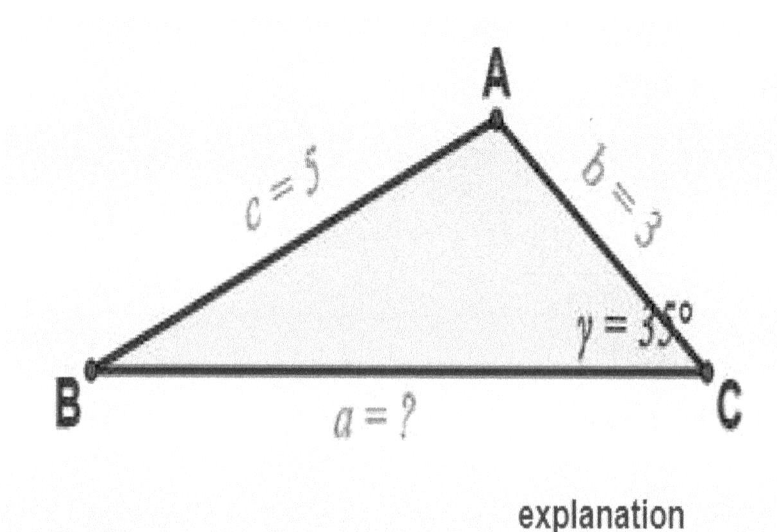

STEP 1: find angle β

To find angle β use Thw Law of Sines:

$$\frac{\sin(\beta)}{b} = \frac{\sin(\gamma)}{c}$$

After substituting $b = 3$, $c = 5$ and $\gamma = 35^o$ we have:

$$\frac{\sin(\beta)}{3} = \frac{\sin(35^o)}{5}$$

$$\frac{\sin(\beta)}{3} = \frac{0.5736}{5}$$

$$\sin(\beta) \cdot 5 = 3 \cdot 0.5736$$

$$\sin(\beta) \cdot 5 = 1.7207$$

$$\sin(\beta) = \frac{1.7207}{5}$$

$$\sin(\beta) = 0.3441$$

$$\beta = \arcsin(0.3441)$$

$$\beta \approx 20.1297^o$$

STEP 2: find angle α

To find angle α use formula:

$$\alpha + \beta + \gamma = 180^o$$

After substituting $\beta = 20.1297^o$ and $\gamma = 35^o$ we have:

$$\alpha + 20.1297^o + 35^o = 180^o$$

$$\alpha + 55.1297^o = 180^o$$

$$\alpha = 180^o - 55.1297^o$$

$$\alpha = 124.8703^o$$

STEP 3: find side a

To find side a use Law of Cosines:

$$a^2 = b^2 + c^2 - 2 \cdot b \cdot c \cdot \cos(\alpha)$$

After substituting $b = 3$, $c = 5$ and $\alpha = 124.8703^o$ we have:

$$a^2 = 3^2 + 5^2 - 2 \cdot 3 \cdot 5 \cdot \cos(124.8703^o)$$

$$a^2 = 9 + 25 - 2 \cdot 3 \cdot 5 \cdot \cos(124.8703^o)$$

$$a^2 = 34 - 2 \cdot 15 \cdot \cos(124.8703^o)$$

$$a^2 = 34 - 30 \cdot (-0.5717)$$

$$a^2 = 34 - (-17.1516)$$

$$a^2 = 51.1516$$

$$a = \sqrt{51.1516}$$

$$a \approx 7.152$$

Peter Chew Method: Using cosine rule

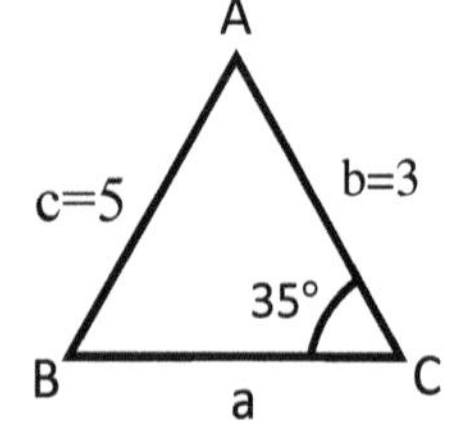

$$5^2 = a^2 + 3^2 - 2(a)(3)\ cos35°$$

$$a^2 - 4.9149\ a\ - 16 = 0$$

$\boldsymbol{a}$ = 7.152, -2.237 (rejected, a > 0)

4. Application of Peter Chew Method for Solution Of Triangle in Civil Engineering

Example 1: The elevation angles of a leaning tower top measure from point B on ground level are **30°**. Given that original completed height of leaning tower (AC) = 73 m and BC= 120 m , find AC [distance leaning tower top to point B].

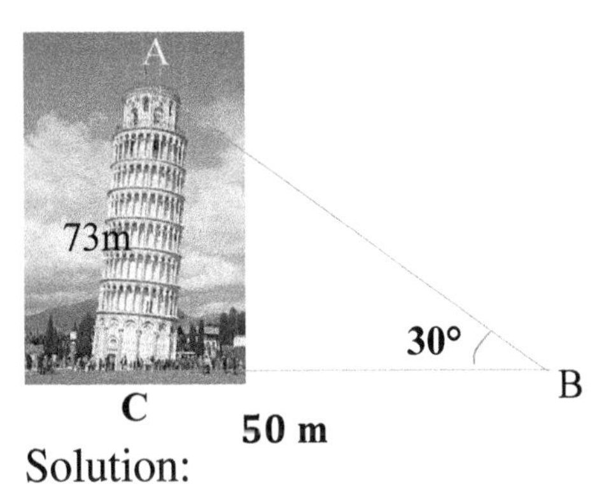

Solution:

Current Method:

Step 1: $\frac{50}{\sin A} = \frac{73}{\sin 30°}$

Rearranging gives us:

$$\sin A = \frac{50 \sin 30°}{73}$$

$$\sin \text{A} = 0.3425$$

A = 20.03° , 159.97° (Reject because 159.97°+ 30° > 180°).

Step 2: When A = 20.03°,

C = 180° - 20.03° ° - 30° = 129.97°

Step 3: When C = 129.97°

$$\frac{c}{\sin 129.97°} = \frac{73}{\sin 30°}$$

$$c = \frac{73\ sin\ 129.97°}{sin\ 30°}$$

= 111.9 m

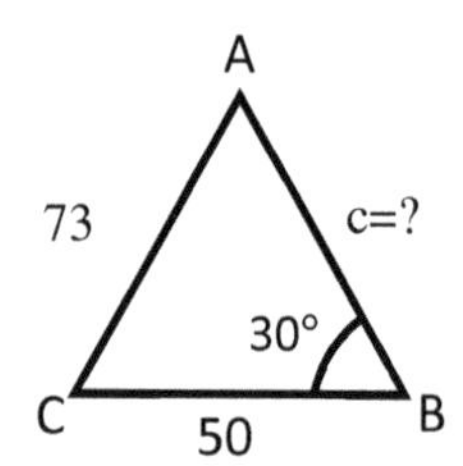

Peter Chew Method: Using cosine rule

$$73^2 = 50^2 + c^2 - 2(50)(c)\ cos30°$$

$$c^2 - 86.60\ c - 2829 = 0$$

c = 111.9 m , -21.28 m(Reject because c > 0).

Example 2: The elevation angles of a leaning tower top measure from point B on ground level are **50°**. Given that original completed height of leaning tower (AC) = 165 m and BC= 120 m , find AC [Distance leaning tower top to point B].

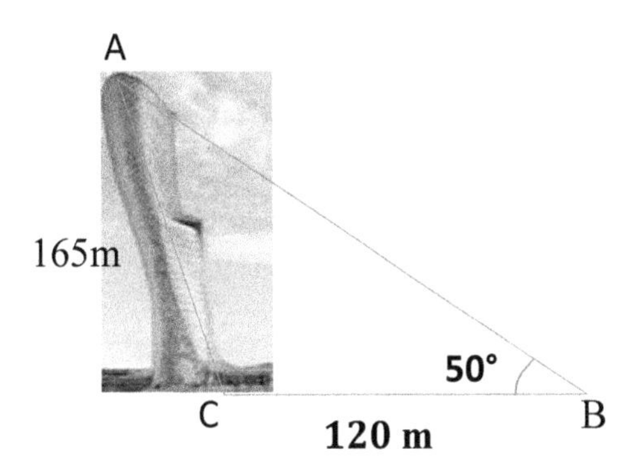

Current Method:

Step 1: $\frac{120}{\sin A} = \frac{165}{\sin 50°}$

Rearranging gives us: $\sin A = \frac{120 \sin 50°}{165}$

$\sin A = 0.5571$

$A = 33.86°$, 146.14° (Reject because 146.14°+ 50°> 180°).

Step 2: When A = 33.86° ,

$C = 180° - 33.86° - 50° = 96.14°$

Step 3: When C = 96.14°

$$\frac{c}{\sin 96.14°} = \frac{165}{\sin 50°}$$

$$c = \frac{165\ sin\ 96.14°}{sin\ 50°}$$

= 214.2 m

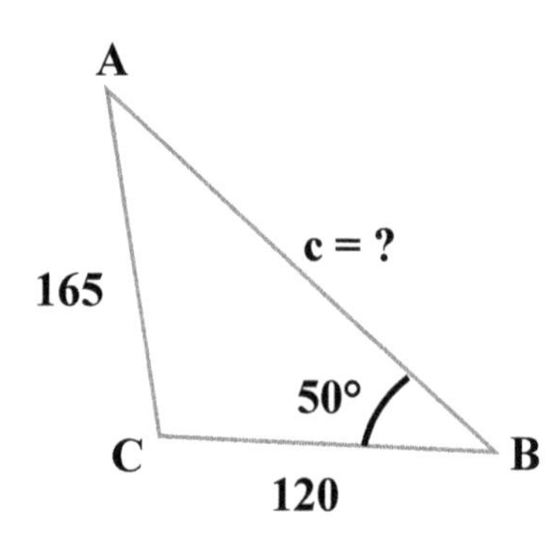

Peter Chew Method: Using cosine rule

$$165^2 = 120^2 + c^2 - 2(120)(c)\ cos50°$$

$$c^2 - 154.3\ c - 12\ 825 = 0$$

c = 214.2 , -59.88(Reject because c > 0).

5. Conclusion

The purpose of Peter Chew's method for solution of triangle is to provide a simple method to compare current methods to aid in mathematics teaching and learning . The application of Peter Chew method for solution of triangle in civil engineering can make the teaching and learning of civil engineering easy.

The purpose of Peter Chew's method for solution of triangle is the same as **Albert Einstein's famous quote Everything should be made as simple as possible, but not simpler.**

In addition, Albert Einstein's also quote :

i)**If you can't explain it simply you don't understand it well enough**,

ii)**We cannot solve our problems with the same thinking we used when we created them**.

iii) *When the solution is simple, God is answering.*

From the Albert Einstein's quote above, it can be seen that simplifying knowledge is very important.

6. Reference

1. Chew, Peter, Peter Chew Method For Solution Of Triangle (August 29, 2022). Available at SSRN: http://dx.doi.org/10.2139/ssrn.4203746

2. Leaning Tower of Pisa. Wikipedia, the free encyclopedia. https://en.wikipedia.org/wiki/Leaning_Tower_of_Pisa .

3. "DiPI Online". *Dizionario di Pronuncia Italiana (in Italian).* Archived *from the original on 30 October 2020.* Retrieved 26 December 2020.

4. "Leaning Tower of Pisa Facts". *Leaning Tower of Pisa.* Archived *from the original on 11 September 2013.* Retrieved 5 October 2013.

5. "Europe | Saving the Leaning Tower". *BBC News. 15 December 2001.* Archived *from the original on 21 September 2013*. Retrieved 9 May 2009.

6. "Tower of Pisa". *Archidose.org. 17 June 2001. Archived from* the original *on 26 June 2009*. Retrieved 9 May 2009.

7. "Leaning Tower of Pisa (tower, Pisa, Italy) – Britannica Online Encyclopedia". *Britannica.com.* Archived *from the original on 8 March 2013*. Retrieved 9 May 2009.

8. "Leaning tower of Pisa loses crooked crown". *Irish News.* Archived *from the original on 28 November 2020.* Retrieved 10 June 2020.

9. Capital Gate. Wikipedia, the free encyclopedia. https://en.wikipedia.org/wiki/Capital_Gate

10. "Capital Gate / RMJM". *ArchDaily. 2018-04-28.* Archived *from the original on 2018-07-22.* Retrieved 2018-09-27.

11. "Capital Gate". *Abu Dhabi National Exhibitions Company (ADNEC). 2010. Archived from* the original *on 11 June 2010*. Retrieved 7 June 2010.

12. Mace Group, http://www.macegroup.com/media-centre/advanced-diagrid-technology-gives-shape-to-capital-gate Archived 2015-10-01 at the Wayback Machine | retrieved=July 29, 2015

13. "Backgrounder - Capital Gate Abu Dhabi". *Hyatt Hotels*. Archived *from the original on 2018-09-09.*

14. Capital Gate Atlas Obscura

(www.atlasosbcura.com). Retrieved on 2019-08-04.

15. Miloš Petrović, Math portal https://www.mathportal.org/calculators/plane-geometry-calculators/sine-cosine-law-calculator.php?combo1=1&val1=&val2=3&val3=5&val4=&val5=&val6=35&val7=

Chapter 3 : Application of Peter Chew Theorem in Civil Engineering

Chew, Peter, Application of Peter Chew Theorem in Civil Engineering (November 22, 2021). Available at SSRN: https://ssrn.com/abstract=3968741 or http://dx.doi.org/10.2139/ssrn.3968741

Abstract. :

Presenting numbers in surd[1] form is quite common in science and engineering especially where a calculator is either not allowed or unavailable, and the calculations to be undertaken involve irrational values.

The Objectives Peter Chew Theore m[2] is to let upcoming generation solve same problem of quadratic roots can solve directly and more easy compare what`s now solution. Therefore, the application of Peter Chew theorem in civil engineering can make the teaching and learning of civil engineering easier.

Keywords: Civil Engineering, Application Peter Chew Theorem, Quadratic Surds.

1. INTRODUCTION

1.1 Presenting numbers in surd form is quite common in Science and Engineering.

Presenting numbers in surd[1] form is quite common in science and engineering especially where a calculator is either not allowed or unavailable, and the calculations to be undertaken involve irrational values.

Common applications of surds include solving a quadratic equation by formula and obtaining the values of trigonometric angles.

Cosine rules:

$i)\ a^2 = b^2 + c^2 - 2bc\ cos\angle A$

$ii)\ b^2 = a^2 + c^2 - 2ac\cos\angle B$

$iii)\ c^2 = a^2 + b^2 - 2ab\ cos\angle C$

1.2 Peter Chew Theorem[2]

The Objective of Peter Chew Theorem is to make it easier and faster to solve the problem of quadratic roots, by converting any value of the Quadratic Surds ($\sqrt{\boldsymbol{a} + \boldsymbol{b}\sqrt{\boldsymbol{c}}}$) into the sum or difference of two real numbers or the sum or difference of two complex number.

Therefore, the application of Peter Chew's theorem in civil engineering can make the teaching and learning of civil engineering easier.

Peter Chew theorem is AI age knowledge because the theorem can help convert all Quadratic Surds ($\sqrt{\boldsymbol{a} + \boldsymbol{b}\sqrt{\boldsymbol{c}}}$) . In addition, the theorem can help convert easier and faster than current method.

This will cause students to increase their interest in using Peter Chew theorem and increase the promotion of effective mathematics learning. When the future epidemics such as Covid-19 occur in the future, it can effectively help mathematics teaching, especially for students studying at home.

1.3 Civil Engineering, The Leaning Tower

i) Leaning Tower of Pisa in 2013

The Leaning Tower of Pisa[3] (Italian: *torre pendente di Pisa*), or simply the Tower of Pisa (*torre di Pisa* [ˈtorre di ˈpiːza; ˈpiːsa][4]), is the *campanile*, or freestanding bell tower, of the cathedral of the Italian city of Pisa, known worldwide for its nearly four-degree lean, the result of an unstable foundation.

The tower is situated behind the Pisa Cathedral and is the third-oldest structure in the city's Cathedral Square (*Piazza del Duomo*), after the cathedral and the Pisa Baptistry.

The height of the tower is 55.86 metres (183 feet 3 inches) from the ground on the low side and 56.67 m (185 ft 11 in) on the high side. The width of the walls at the base is 2.44 m (8 ft 0 in).

Its weight is estimated at 14,500 tonnes (16,000 short tons).[5] The tower has 296 or 294 steps; the seventh floor has two fewer steps on the north-facing staircase.

The tower began to lean during construction in the 12th century, due to soft ground which could not properly support the structure's weight, and it worsened through the completion of construction in the 14th century.

By 1990, the tilt had reached 5.5 degrees.[6,7,8] The structure was stabilized by remedial work between 1993 and 2001, which reduced the tilt to 3.97 degrees.[9]

ii) Capital Gate in 2013

Capital Gate[10], also known as the Leaning Tower of Abu Dhabi, is a skyscraper in Abu Dhabi that is over 160 meters (520 ft) tall, 35 stories high, with over 16,000 square meters (170,000 sq ft) of usable office space.[11]

Capital Gate is one of the tallest buildings in the city and was designed to incline 18° west.[12] The building is owned and was developed by the Abu Dhabi National Exhibitions Company. The tower is the focal point of Capital Centre.

Foundation

The structure rests on a foundation of 490 pilings that have been drilled 30 meters (98 ft) below ground. The deep pilings provide stability against strong winds, gravitational pull, and seismic pressures that arise due to the incline of the building.

Of the 490 pilings, 287 are 1 meter (3 ft 3 in) in diameter and 20 to 30 meters (66 to 98 ft) deep, and 203 are 60 centimeters (24 in) in diameter and 20 meters (66 ft) deep.

All 490 piles are capped together using a densely reinforced concrete mat footing nearly 2 meters (6.6 ft) deep. Some of the piles were only initially compressed during construction to support the lower floors of the building. Now they are in tension as additional stress caused by the overhang has been applied.[13]

Architecture and design

The building has a diagrid specially designed to absorb and channel the forces created by wind and seismic loading, as well as the gradient of Capital Gate.

Capital Gate is one of only a handful of diagrid buildings in the world. Others include London's 30 St Mary Axe (Gherkin), New York's Hearst Tower, and Beijing's National Stadium.[13]

Capital Gate was designed by architectural firm RMJM and was completed in 2011. The tower includes 16,000 square meters (170,000 sq ft) of office space and the Andaz Hotel on floors 18 through 33.[14,15]

2. Application of Peter Chew Theorem in Civil Engineering

Civil Engineering (The original completed height of leaning tower)

Example 1: The elevation angles of a leaning tower top measure from two point D and B on ground level are 45° and 60° respectively. Given that that DB = $(2\sqrt{3} + 4\ )$ m, BC = $(4\sqrt{3} + 2\ )$ m, find the original completed height of leaning tower (AC). Figure 4.

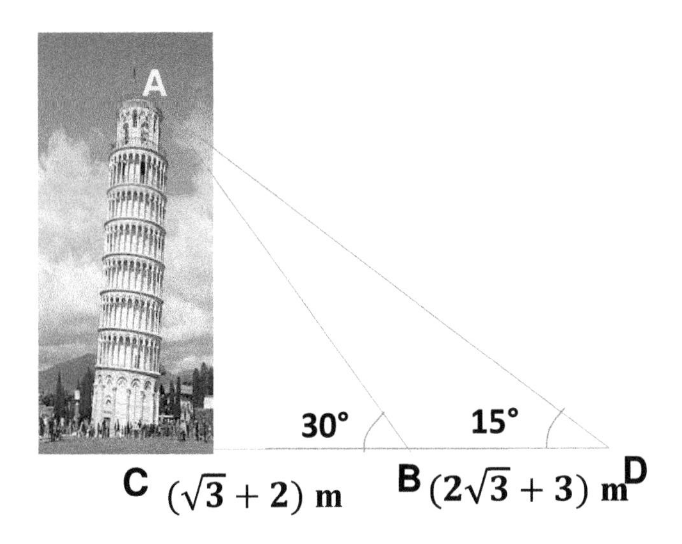

Figure 4

Solution:

$\angle BAD = 30° - 15° = 15°$

$AB = (2\sqrt{3} + 3)$ m. [$\angle BAD = \angle BDA$]

Cosine rule,

$$AC^2 = a^2 + c^2 - 2ac\cos\angle B$$

$$AC^2 = (\sqrt{3} + 2)^2 + (2\sqrt{3} + 3)^2 - 2(\sqrt{3} + 2)(2\sqrt{3} + 3)\cos 30°$$

$$= 3 + 4 + 4\sqrt{3} + 12 + 9 + 12\sqrt{3} - 2(6 + 3\sqrt{3} + 4\sqrt{3} + 6)\cos 30°$$

$$= 28 + 16\sqrt{3} - (21 + 12\sqrt{3})$$

$$= 7 + 2\sqrt{12}$$

$$AC = \sqrt{7 + 2\sqrt{12}}$$

Current Method,

Let $\sqrt{7+2\sqrt{12}}$ be $\sqrt{x}+\sqrt{y}$

$$7+2\sqrt{12} = (\sqrt{x}+\sqrt{y})^2$$

$$= x + y + 2\sqrt{xy}$$

Comparing the two sides of the above equation, we have x + y = 7

y = 7 - x ..... i)

And xy = 12 ....ii)

Substitute i) in ii),

$$x(7-x) = 12$$

$$x^2 - 7x + 12 = 0$$

$$(x-4)(x-3) = 0$$

$$x = 4, 3$$

From i), when x =4, y = 7- 4 = 3

When x =3, y = 7- 3 = 4

$AC = \sqrt{4} + \sqrt{3}$

$= 2 + \sqrt{3}$

Original completed height of leaning tower (AC) is $(2 + \sqrt{3})$m .

Peter Chew Theorem,

Cause x^2 - 7x + 12 = 0, then x = 4 , 3

$\therefore AC = \sqrt{4} + \sqrt{3}$

$= 2 + \sqrt{3}$

Original completed height of leaning tower (AC) is $(2 + \sqrt{3})$ m

Example 2: The elevation angles of a leaning tower top measure from two point D and B on ground level are 30° and 60° respectively. Given that that DB = $(8\sqrt{5}+24)$ m, BC = $(8\sqrt{5}+8)$ m, find the original completed height of leaning tower (AC). Figure 4.

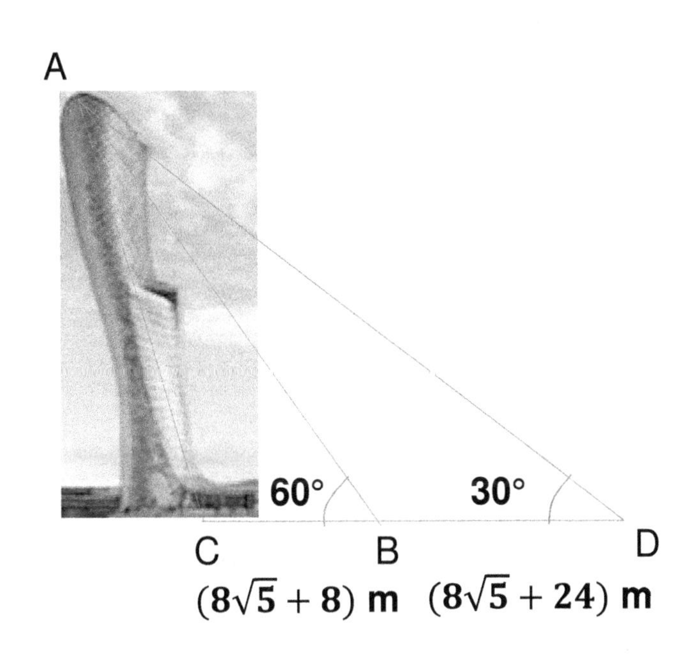

Figure 4

Solution:

$\angle BAD = 60° - 30° = 30°$

$AB = (8\sqrt{5}+24)$ m. [$\angle BAD = \angle BDA$]

Cosine rule,

$AC^2 = a^2 + c^2 - 2ac\ cos\angle B$

$AC^2 = \left(8\sqrt{5} + 24\right)^2 + \left(8\sqrt{5} + 8\right)^2$

$- 2\left(8\sqrt{5} + 24\right)\left(8\sqrt{5} + 8\right) \cos 60°$

$= 320 + 576 + 384\sqrt{5} + 320 + 64 + 128\sqrt{5}$

$- (320 + 64\sqrt{5} + 192\sqrt{5} + 192)$

$= (768 + 256\sqrt{5})$

$= (768 + 2\sqrt{81920})$

$AC = \sqrt{768 + 2\sqrt{81920}}$

Current Method,

Let $\sqrt{768 + 2\sqrt{81920}}$ **be** $\sqrt{x} + \sqrt{y}$

$768 + 2\sqrt{81920} = (\sqrt{x} + \sqrt{y})^2$

$= x + y + 2\sqrt{xy}$

Comparing the two sides of the above equation,

we have $x + y = 768$

$y = 768 - x$ **….. i)**

And $xy = 81920$ **……..ii)**

Substitute i) in ii), $x (768 - x) = 81920$

$$x^2 - 768x + 81920 = 0$$

$$(x - 640)(x - 218) = 0$$

$$x = 640,128$$

From i), when x =640, y = 768- 648 = 128

When x =218, y = 864- 128 = 640

$$AC = \sqrt{640} + \sqrt{128}$$

$$= 8\sqrt{10} + 8\sqrt{2}$$

Original completed height of leaning tower (AC) is $(8\sqrt{10} + 8\sqrt{2})\,m$.

Peter Chew Theorem,

Cause x^2 -768x +81920 = 0, then x=640,218

$$\therefore AC = \sqrt{640} + \sqrt{128}$$

$$= 8\sqrt{10} + 8\sqrt{2}$$

Original completed height of leaning tower (AC) is $(8\sqrt{10} + 8\sqrt{2})m$.

3. Conclusion

The application of Peter Chew theorem in civil engineering can make the teaching and learning of civil engineering easier.

4. Reference

[1]. Shefiu S. Zakariyah, PhD Surds Explained with Worked Examples. (26, 30) Feb.2014.
https://www.academia.edu/6086823/Surds_Explained_with_Worked_Examples

[2]. Peter Chew . Peter Chew Theorem and Application.
Chew, Peter, Peter Chew Theorem and Application (March 5, 2021). Available at SSRN: https://ssrn.com/abstract=3798498 or http://dx.doi.org/10.2139/ssrn.3798498 or Europe PMC: PPR: PPR300039

[3]. Leaning Tower of Pisa. Wikipedia, the free encyclopedia.
https://en.wikipedia.org/wiki/Leaning_Tower_of_Pisa

[4]. "DiPI Online". *Dizionario di Pronuncia Italiana (in Italian).* Archived *from the original on 30 October 2020.* Retrieved 26 December 2020.

[5]. "Leaning Tower of Pisa Facts". *Leaning Tower of Pisa.* Archived *from the original on 11 September 2013.* Retrieved 5 October 2013.

[6]. "Europe | Saving the Leaning Tower". *BBC News. 15 December 2001.* Archived *from the original on 21 September 2013*. Retrieved 9 May 2009.^

[7]. "Tower of Pisa". *Archidose.org. 17 June 2001. Archived from* the original *on 26 June 2009*. Retrieved 9 May 2009.

[8]. "Leaning Tower of Pisa (tower, Pisa, Italy) – Britannica Online Encyclopedia". *Britannica.com.* Archived *from the original on 8 March 2013*. Retrieved 9 May 2009.

[9]. "Leaning tower of Pisa loses crooked crown". *Irish News.* Archived *from the original on 28 November 2020*. Retrieved 10 June 2020.

[10]. Capital Gate. Wikipedia, the free encyclopedia. https://en.wikipedia.org/wiki/Capital_Gate

[11]. "Capital Gate / RMJM". *ArchDaily. 2018 04 28.* Archived *from the original on 2018-07-22*. Retrieved 2018-09-27.

[12]. "Capital Gate". *Abu Dhabi National Exhibitions Company (ADNEC). 2010. Archived from* the original *on 11 June 2010*. Retrieved 7 June 2010.

[13]. Mace Group, http://www.macegroup.com/media-centre/advanced-diagrid-technology-gives-shape-to-capital-gate Archived 2015-10-01 at the Wayback Machine | retrieved=July 29, 2015

[14]. "Backgrounder - Capital Gate Abu Dhabi". *Hyatt Hotels.* Archived *from the original on 2018-09-09.*

[15]. Capital Gate Atlas Obscura

(www.atlasosbcura.com). Retrieved on 2019-08-04.

Ingram Content Group UK Ltd.
Milton Keynes UK
UKHW020834210723
425555UK00014B/542